California Natural History Guides: 6

WEATHER

OF THE

SAN FRANCISCO BAY REGION

BY

HAROLD GILLIAM

DRAWINGS BY GENE CHRISTMAN

UNIVERSITY OF CALIFORNIA PRESS

BERKELEY, LOS ANGELES, LONDON

UNIVERSITY OF CALIFORNIA PRESS
BERKELEY AND LOS ANGELES, CALIFORNIA
UNIVERSITY OF CALIFORNIA PRESS, LTD.
LONDON, ENGLAND
© 1962 BY THE REGENTS OF THE UNIVERSITY OF CALIFORNIA

ISBN: 0-520-00469-8
LIBRARY OF CONGRESS CATALOG CARD NUMBER: 62-17533
PRINTED IN THE UNITED STATES OF AMERICA

8 9 0

CONTENTS

ILLUSTRATION ON COVER
    Fog rolling into Sausalito

ACKNOWLEDGMENT

The author wishes to express his appreciation to C. Robert Elford, U. S. Weather Bureau State Climatologist, who gave invaluable technical assistance in the preparation of this book.

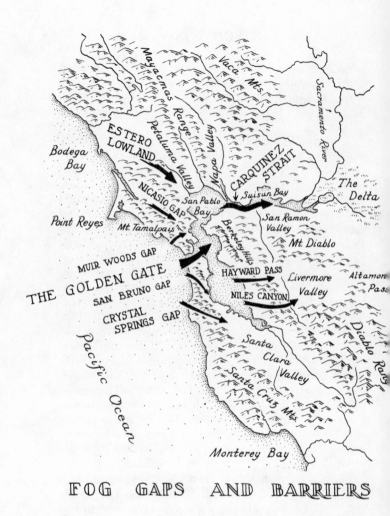

FOG GAPS AND BARRIERS

INTRODUCTION

"If you don't like the weather," they say in New England, "wait a few minutes."

In the San Francisco Bay Region the advice might be amended: "If you don't like the weather, walk a few blocks."

A single block, in fact, might be enough. In certain districts of San Francisco a house in one block might be wrapped in fog while residents of a house a block or so away, behind a protecting spur of one of the city's hills, may sunbathe in perfect comfort in their own back yard.

At greater distances within the ten-county Bay Region, the weather differences increase proportionately. Fishermen along the fog-shrouded coast of Marin County on a summer day may be shivering in the low fifties while people in San Rafael, ten miles east, bask in comfortable 70-degree weather, and residents of ranches at the edge of the Sacramento Valley, another 40 miles east, mop their brows as the thermometer hits 100—a temperature difference of 50 degrees in 50 miles.

During the winter there are similar differences. Boulder Creek, in the Santa Cruz Mountains, averages 60 inches of rain a year, while just over the hills in the Santa Clara Valley the annual total is only 13 inches.

Man has not yet fulfilled his age-old dream of controlling the weather, but in the Bay Region he comes close; he can change the weather around him by moving a short distance. Probably no comparable area on earth displays as many varieties of weather simultaneously as the region around San Francisco Bay.

The Weather Funnel

The reasons for this unique situation lie in California's extraordinary geography. The prime movers in setting up the geography were two mountain ranges, a river system, and a big thaw.

[5]

The Sierra Nevada, a granite barrier rising from 10,000 to 14,000 feet into the sky some 200 miles inland from the shore of the Pacific, intercepts the clouds and moisture-laden winds drifting eastward from the ocean and forces them to drop their burden on the mountain slopes in the form of rain and snow.

The water, cascading down the western slope of the Sierra in a vast network of creeks, waterfalls, streams, and rivers, merges in the Central Valley to form the greatest river system within the boundaries of any single state. This tremendous volume of water, slicing through the Coast Range to the sea, carved the Carquinez Strait and the Golden Gate long before San Francisco Bay was formed.

At the end of the last Ice Age, the great glaciers melted in such volume that the oceans overflowed. Over a period of thousands of years, rising seas flooded through the river-carved gorge at the Golden Gate and occupied an inland valley to create San Francisco Bay. The river no longer flowed to the sea but emptied into the bay at Carquinez Strait.

Thus the successive action of the river and the ocean created the only complete breach in the Coast Range, which borders the Pacific for most of California's length. As a result, the San Francisco Bay Region is the meeting place of continental and oceanic air masses. Through the funnel of the Golden Gate and San Francisco Bay, the immense aerial forces of sea and land wage a continual war, and the tide of battle often flows back and forth with regularity. The line between the two types of air masses, particularly in summer, may zigzag through the streets of San Francisco and extend in similar erratic fashion across the entire region.

THE SUBDIVIDED RANGE

The reasons for the zigzags—the highly variable weather patterns within the region—is the complex topography of the Coast Range, which modifies in in-

[6]

tricate ways the basic struggle between air masses of land and sea. The weather of any single mountain range, with its ridges and canyons and valleys, is complex enough, but the section of the Coast Range comprising the Bay Region divides and subdivides into various sub-ranges, each with its own hill-and-valley topography creating its own modifications of the basic weather and climate patterns. (Climate is simply average weather over a period of time.)

In general the Coast Range in this region is a double chain of mountains running north and south (or, more precisely, north-northwest and south-southeast). Between the two chains lies the basin of San Francisco Bay, including the valleys at the ends of the bay: Petaluma on the north and Santa Clara on the south.

The western range consists of the Santa Cruz Mountains south of the Golden Gate, and the Marin hills, including Mount Tamalpais, to the north. As if to complicate matters further, the eastern part of the Coast Range in this vicinity is itself divided into two main chains. Immediately to the east of the bay are the Berkeley Hills, paralleled, beyond the San Ramon and Livermore valleys, by the higher Diablo Range. North of the bay this double aspect of the range continues but is further subdivided into subsidiary chains, including the Sonoma, the Mayacmas, and the Vaca mountains.

MICROCLIMATES

Eastward from the ocean over the several ranges, each successive valley has less of a damp, seacoast climate and more of a dry, continental climate—hotter in summer and colder in winter. But this basic pattern is further modified and complicated by a number of gaps and passes in the ranges, most important of which is the Golden Gate, allowing the easy penetration of seacoast weather inland.

The pattern is also modified by large bodies of water, which tend to cool their shores in the summer and warm

[7]

them in the winter. The most important of these, of course, is San Francisco Bay itself, and its various subdivisions and tributaries, including San Pablo Bay, Suisun Bay, and the Delta, where the major rivers of the Sierra and the Central Valley meet in an intricate network of watercourses and low islands.

Because of these complex forms of the land, there is actually no such thing as Bay Region climate. There are only innumerable microclimates within the region, varying widely from mountain to mountain, from valley to valley, and from point to point within the mountains and valleys.

The results are manifold: the great flowing fogs that move through the Golden Gate from the ocean in the summer; the warm dry winds that whip down through the canyons to the bay in spring and fall, scuffing the bay surface into whitecaps; the cumulus clouds that drift eastward across the sky in winter, throwing moving patterns of light and shade across land and water; the massive cumulonimbus clouds that build up to heights of several thousand feet above the rim of mountains around the bay; the rains that deluge one valley while scarcely dampening the next; the snows that occasionally dust the tops of the highest peaks; and the frosts that descend to the lowlands.

Besides the city dweller who must merely decide whether or not to take a coat to the office in the morning, thousands of people are greatly concerned with all these phenomena—people whose work is more directly affected—pilots of ships and small boats who must learn to navigate blindly in the fogs that hug the waters and to cope with churning seas stirred up by winds; linemen who maintain power lines and telephone wires damaged by storms; road workers who must keep the streets and highways clear of debris in heavy rains and winds; passengers and crews of the hundreds of planes whose schedules may be upset by fog; carpenters and construction men whose work is

[8]

halted by storms; dairymen who must spend extra money for hay if insufficient rain fails to raise a good crop of grass; painters on the bay's great bridges; grape growers in Napa and Sonoma and Livermore who fear late frosts and early rains; orchardists in the Santa Clara Valley who must reckon when to prune and irrigate and plow by observing the signs in the sky.

For the three million people who live and work in the San Francisco Bay Region the weather is the lowest common denominator, consistently the most recurrent single topic of conversation on street corners, in corridors and elevators, in taverns, on buses and trains, and wherever people gather to talk. Although weather talk is common to all mankind, it achieves here a particular flavor and intensity unknown elsewhere.

Residents of other parts of the country are apt to open the conversation with such a remark as: "Some weather we're having lately, isn't it?" But in San Francisco no one—except tourists and newcomers—assumes that the listener has been experiencing the same kind of weather as the speaker, and the opening gambit is likely to be: "What's the weather like in your part of town?" Commuters into San Francisco compare notes on their respective communities, and no one is greatly surprised if Berkeley has fog while Alameda is in bright sun, or Mill Valley has rain while Palo Alto is dry and clear.

The Ocean of Air. Before getting into the detailed peculiarities of the Bay Region's microclimates, we should review some general facts about the atmosphere that apply to all weather and climate everywhere on earth.

1. The ocean of atmosphere that surrounds the earth bears down on the earth's surface with an average "weight" of 14.7 pounds per square inch at sea level. The exact "weight" or air pressure, however, depends on the altitude, air temperature, and other variables.

2. Warm air is light and rises; cold air is heavy and

descends. Because the cold air presses down more heavily on the earth's surface than warm air, a cold region is a relatively high pressure area; a warm region, a low pressure area.

3. Just as water tends to seek its own level, so air tends to equalize its pressure. Thus, air moves from a high pressure area to a low pressure area; winds blow from a cool to a warm area, just as cold air will move through a door or window into a warm room, creating a draft.

4. When air rises, in general it expands and cools at the rate of about 5½ degrees Fahrenheit per thousand feet of elevation.

5. When air descends, it compresses and grows warmer at about the same rate.

6. Warm air is able to hold more moisture (in vapor form) than cool air. If warm, damp air begins to cool off, it will reach a point where it can no longer contain its watery load and the moisture will condense into fog or clouds. If the air continues to grow cooler, the fog or clouds will drop precipitation as rain or snow. A familiar example is the condensation that occurs when air is cooled by contact with a glass of cold water and the condensed moisture is deposited on the outside of the glass.

7. Air moving freely across the surface of the earth tends to curve to the right, clockwise, in the Northern Hemisphere, and to the left in the Southern Hemisphere. This tendency is known as the Coriolis force or Coriolis effect—named for the French scientist who formulated the principle.

To illustrate the Coriolis force, draw on a piece of paper a large circle with a dot in the center. The circle represents the earth as seen from above the North Pole, and the dot is the pole itself. Then rotate the paper slowly in a counterclockwise direction, representing the rotation of the earth. As the paper rotates, draw a short line inside the circle from any point toward some fixed point off the page, such as the wall of the room. You will find that the line on the paper curves to the right.

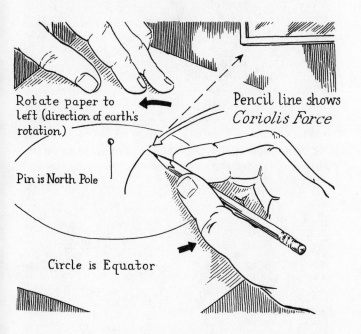

Rotate paper to left (direction of earth's rotation)

Pencil line shows *Coriolis Force*

Pin is North Pole

Circle is Equator

In the same fashion, anything moving freely across the surface of the Northern Hemisphere curves to the right because of the rotation of the earth beneath it. (In the Southern Hemisphere the curve is to the left.) This is true not only of winds but of ocean currents, rockets, and artillery projectiles. Even rifles must be compensated to counteract the rightward drift of the bullet. The rightward drift of winds and ocean currents is particularly important in understanding the Bay Region's weather.

THE FOUR SEASONS

SPRING

During his stay in San Francisco in the 1880's, Robert Louis Stevenson wrote of the "trade winds" that blow in through the Golden Gate and over the hills of the city. The term conveys the proper romantic flavor and conjures visions of fabled galleons on the trade routes to Cathay, but the fact is that there are no trade winds within thousands of miles of San Francisco. Countless writers, attempting to lend a note of glamour to the prevailing sea breeze that blows from the ocean across the city's hills, have repeated Stevenson's error.

The trade winds blow from the east in the latitudes on either side of the Equator. Spanish galleons of the 16th and 17th centuries made use of the Pacific trades. They sailed south from Acapulco to take advantage of the winds that bore them west across the Pacific to Manila. On the return trip, however, they sailed far to the north in order to pick up the prevailing westerlies, a river of air moving from the west across the ocean in the latitude of the United States. The westerlies brought them to the coast of California and then southwest again to Mexico.

It is the westerlies that assault the California coastline, throwing sheets of ocean spray against the headlands, blowing sand along the beaches into great dunes and drifts, flailing the grasses of the coastal slopes, bending the laurels and cypresses into permanently contorted shapes.

In the spring the westerlies are greatly intensified. During March, April, and May, as the days grow longer, the northward-moving sun heats the continent day by day, melting mountain snows, causing flowers to bloom and fruit to bud across foothills and plains and deserts. In the great Central Valley of California, a 500-mile-long basin surrounded by mountains, the heated air rises; the surface pressure, as recorded on barometers, falls; and Weather Bureau forecasters mark the valley on their maps with a large "L" for low pressure. Out over the Pacific, however, the air in contact with the ocean is still cool. It presses down heavily on the water surface; ships relay the barometer readings to the Weather Bureau; and the weather men indicate the high pressure on their maps with the letter "H." The result of this differing pressure is what the meteorologists call an "onshore pressure gradient." The air rushes from the high pressure areas toward the low, from the cool ocean toward the warm inner valleys.

The Pacific High. The onshore winds are also intensified in the spring by the Pacific High, a mass of air that presses down on the ocean a thousand miles or so offshore. Weathermen wish they knew more about its origin and behavior, but in general it is the result of a global process. Air warmed by the hot sun over the Equator rises and heads northward for the Arctic regions. Some of it cools off and sinks to the ocean surface again several thousand miles to the north as the Pacific High—a "mountain" of cool air weighing heavily on the water.

Marked by a big "H" on the weather maps, the Pacific High usually occupies a position somewhere between San Francisco and the Hawaiian Islands, sending out winds in all directions across the ocean surface. Because it is a result of the sun's heating, it usually follows the sun to the south in the winter and migrates northward with the sun in the spring. (Of course, the sun does not literally move north. It appears to do so as the planet swings around to the "springtime" position in its orbit and the tilt of the earth's axis exposes the Northern Hemisphere more directly to the sun's rays.)

For reasons that are not yet clear, the behavior of the Pacific High is erratic. Sometimes it is weak, sending out only gentle breezes; sometimes it is strong, giving rise to great rushing winds. Some years it does not come as far north as usual; on other occasions it moves north-ward far beyond its "normal" location. All these vari-ations in the Pacific High cause "unusual" weather around San Francisco Bay.

In March and April the Bay Region weather is highly variable. Days marked by the last gusts of winter blow-ing themselves out are followed by days of hazy, balmy warmth. The green hills, the grassy vacant lots, and the residents of the region themselves gratefully absorb the increasing sunlight. A light haze hangs over the bay and the hills as moisture is evaporated into the air from the water and from the earth, often still damp from the last showers of winter. Along the ocean boundaries of the land, the battering waves of winter die out and are re-placed by gentle swells.

The surface of San Francisco Bay is glassy. By many an afternoon in early April, the sun has sufficiently heated the land to cause the surface air to warm and rise, creating an area of slightly low pressure. Cooler air from the higher pressure areas offshore moves in from the ocean, ruffles the bay, and clears away the haze.

Meanwhile, as if with a far-off roll of drums, the forces of the atmosphere are being marshaled for a major change. The Pacific High is moving northward to latitudes a thousand miles or so west of San Francisco. Pushed from the rear by the Pacific High and pulled from the front by the heating of the Central Valley and other warming inland regions, masses of air begin mov-ing toward the continent with increasing speed. But like everything else moving on the surface of the earth, the winds are affected by the Coriolis force; they begin to curve to the right. When they reach the edge of the continent they are coming from the northwest, in some places moving almost parallel to California's slanting coastline.

There they are prevented from moving inland in force by the high wall of the Coast Range, and so they blow on down the coast, breaching the mountain barrier only where gaps and passes in the range permit tongues of ocean air to penetrate inland.

Ocean River. Through March and April and May the Pacific High moves closer and grows stronger; the valley continues to warm up. In any latitude ocean currents are created and borne along by the action of the wind. The effect of the northwest wind down the California coast is to push the surface of the ocean before it and create a strong current running southward down the shoreline like a river. But once again the picture is changed by the Coriolis force. Like the winds themselves, the waters, too, tend to curve to the right across the surface of the rotating sphere. The southward-moving current veers offshore at about a 45-degree angle. In order to replace the surface currents moving away from the coast, masses of water surge up from the bottom of the ocean, creating a continual fountain of upwelling waters.

This bottom water, coming from depths of several hundred feet, is often 10 to 15 degrees colder than the sun-warmed surface. Swimmers along northern California beaches are painfully aware that these waters are frigid—often in the low 50's—compared to the water temperatures of the surf along the beaches of southern California, which are beyond the principal zone of upwelling and are frequently in the high 60's. Off Cape Mendocino, for example, 200 miles north of the Golden Gate, the spring and summer upwelling reduces the temperature of the water, making it even colder than in winter.

The Great Fog Bank. This streak of cold water along the coast is a basic part of the Bay Region's summertime air-conditioning system. The wind from the west, having traversed thousands of miles of ocean, absorbs great quantities of moisture from the surface by evaporation. When it approaches the coast, the air comes into contact

[15]

with the cold, upwelled waters and is itself cooled off, causing its moisture to condense. The same process that causes moisture to form on the outside of a glass of cold water takes place here on a mammoth scale. Rather than clinging to the outside of a glass, the drops of water condensed in the sea breeze cling to the minute particles of salt that have been thrown into the air with the spray. Thus, as the wind blows over the cold surface, the drops of water continue to form until they create a haze, which soon thickens into fog.

Thus is formed the great fog bank that hangs along most of the California coast intermittently during the late spring and the summer. It may range in width from a hundred yards to more than a hundred miles, in height from a hundred feet to half a mile. In the beginning, however, it takes form merely as puffs and wisps of vapor that drifts landward and cling to the shoreward slopes of the coastal hills or hang over the beaches. Then, as the season advances and the wind increases, the entire process accelerates; the separate vapors merge into a solid mass that steals up the coastal canyons in late afternoon or early evening when the sun's rays are no longer strong enough to burn it off. It moves into the upper canyons, where it collects on the branches of the redwoods and falls to the ground, keeping the earth damp. The big trees thrive on this moisture. Redwood country is fog country; the *Sequoia sempervirens* seldom grows beyond the range of the sea fog.

Along most of the Coast Range the sea air and its fog reach the heads of the canyons and are stopped by the higher ridges from penetrating farther inland. But at the Golden Gate, the only sea-level breach in the mountains, the wind moves through the range, bringing with it the masses of condensed moisture. The fog is first visible in the Golden Gate in early spring, perhaps appearing one April morning as a wispy finger of vapor entering the bay beneath the 230-foot-high deck of the Golden Gate Bridge, heralded by the sonorous trumpeting of the fog horns. In the course of a few hours it may thicken until it is a solid mass moving through the mile-wide strait into the bay. At the maximum, perhaps a million gallons of water an hour float through the Gate as vapor.

The moving wall of vapor envelops the deck of the bridge and swallows ships and shores as it advances. But the fog itself is not the primary element at work. It is created by and borne upon the wind. The wind is often present when there is no fog. When the wind is sufficiently laden with moisture and sufficiently cooled by the water, the fog comes into being and is carried inland by the moving air. Thus the wind strikes a par-ticular area in advance of the fog.

The fog formed in this way is literally a cloud resting on the water. It may later rise into the air and look more like the conventional idea of a cloud, but it does not change except in elevation. Visitors are often puzzled when Californians refer to this cloud layer as a "high fog," but the terms are interchangeable. Weather men call it "stratus," which is one type of cloud (see page 63), whether it is at ground level or overhead. Stratus and fog are identical.

Spring is the best time to observe the fantastic fog forms that develop as the vapors encounter obstacles and create such shapes as wreaths and domes over Alca-traz and Angel islands; arches over the ridge at the Marin end of the bridge; eddies, falls, and cascades over the hills immediately north of the Gate, down the east-ern slopes of Twin Peaks and the hills of the Peninsula; surges and combers as seen from the top of Tamalpais; and fog decks that form near the top of the Berkeley Hills and build out toward the bay.

Sometimes fog is formed suddenly by very odd cir-cumstances. At times when the damp ocean air is al-most at the saturation point but condensation has not yet occurred, anything that lowers the temperature a few degrees can cause condensation to take place. A high layer of clouds moving in front of the sun and throwing the surface into shadow can cause sufficient cooling of the air for the stratus to form. An eclipse of the sun has been known to create a sudden fog lasting for the duration of the eclipse. Nearly saturated air moving across a small body of water, such as Stow Lake in Golden Gate Park, will drop in temperature sufficiently on contacting the slightly cooler water to form small wisps of fog that drift across the lake, duplicating in miniature the stratus formations over the ocean.

A similar phenomenon on a larger scale takes place in the Golden Gate when clear air near the saturation point is cooled a degree or two by a reversal of the tidal current beneath it. When the tide changes from ebb to flood, the warm outgoing waters of the bay and its tributary rivers are replaced by the cold incoming waters of the ocean. The cold water lowers the air temperature enough for the vapor to condense into fog. Similarly, a tidal shift from flood to ebb may cause the fog to disappear.

The spring and summer fogs of the Bay Region are known as "advection" fogs. Just as "convection" refers to a vertical movement of air or liquid (as takes place when water boils), advection is a horizontal movement, and an advection fog is formed by the horizontal movement of the air.

At times when the wind is not damp enough or the ocean cold enough to form stratus over the water, the formation may take place as the wind hits the coastal hills and is forced to rise. As it rises, it cools sufficiently to cause its vapor to condense in a fog deck against the hills or perhaps in isolated puffs and wisps which are then carried inland by the wind and float like big chunks of cotton at rooftop level across downtown San Francisco.

SUMMER

Week by week through March and April and into June the forces that produce the stratus increase in intensity. The Pacific High moves farther north, closer to the latitude of San Francisco, sending out stronger winds; offshore the upwelling of the cold, bottom waters increases, condensing the winds' moisture into thicker masses of fog; in the Central Valley the northward-moving sun sends temperatures to the 100 mark and beyond.

Along the valley floor, in orchards, vineyards, and cotton fields, the crops ripen in the heat. The hot air rises, sucking cool masses in great drafts through the only break in the valley's surrounding mountains, San Francisco Bay. With the ocean air comes the fog, evaporating gradually in the hot dry air of the valley, but sometimes penetrating at night as far as Sacramento and Stockton.

Over a period of days the draft from the bay has its effect on the valley. The incoming, cool, heavy sea air begins to replace the rising, warm land air, and the valley, particularly the part nearest the bay, cools off. Without the intense valley heat to suck the sea air in through the Bay Region, the wind diminishes and no longer carries the stratus inland. When the valley cools sufficiently, the entire fog-producing system will break down. San Francisco, the Golden Gate, and the coastline will be fog free.

Then the whole process starts over again. With the wind and fog gone, the sun gradually reheats the valley. The rising warm air once again begins to suck the marine air inland from the ocean, through the Golden Gate and across the bay, producing the stratus once more.

Multicycles. Thus, the fog over the bay normally ebbs and flows in cycles. There are at least three kinds of cycles. The cycle described above may take place over a period of from three days to several weeks, but it averages about a week in length and can be called a weekly cycle. Even more obvious is the daily cycle. In the afternoons the stratus moves in through the Golden Gate and other gaps in the coastal hills, spreads over a large area of the bay and its shores during the cool hours of the night, then is burned off by the morning sun.

The daily cycle takes place in this particular way only at the early stages of the weekly cycle when there is an equilibrium of the fog-producing forces. Later, when the stratus covers the coast and the central Bay Region for days at a time, the daily cycle is more evident farther inland. San Francisco and the immediate vicinity may be blanketed for several successive days, while Palo Alto to the south and San Rafael to the north are beyond the fog's reach. At night, however, the white mists may roll in over the coastal hills, engulfing everything from the Santa Clara Valley to the Napa Valley and as far inland as Stockton, only to be burned off in these outer areas by the morning sun.

The third type of cycle is seasonal; it begins in early spring. As the Pacific High grows stronger and the valley heat increases, the stratus penetrates farther inland and remains longer with each weekly cycle, reaching a maximum in August, then decreasing through September.

Finally, there are long-term tendencies that may be too irregular to be called cycles and as yet have not been sufficiently studied to be fully explained. Several successive summers with very little fog may be followed by several very foggy summers. On the other hand, a foggy summer may be succeeded the next year by a very clear summer.

All the cycles—daily, weekly, seasonal, and long-term —as described here, are ideal or theoretical conditions which are only approximated in varying degrees by the actual conditions at any particular time. Often the stratus comes and goes erratically without apparent obedience to the "rules." These variations result from complex changes in the Pacific High, the prevailing westerlies, the ocean currents, and the heating of the land by the sun.

The Inversion. Although the advection fog is formed primarily by the horizontal movement of air, in the development of the fog formations and cycles, vertical movements are important as well. As the cool, heavy ocean air moves in across the bay and its shores, it slides beneath the warmer air. Thus there is a layer of cool, moist air at the surface, topped by warmer air above.

Because the more normal condition of the atmosphere around the world is that air is warm at the surface and becomes cooler at higher elevations, this peculiar situation over the Bay Region is an "inversion"—the usual atmospheric condition is inverted.

Anyone who climbs the hills around the bay in midsummer may be aware of sharp changes in temperature that come with changes in elevation. At a certain height above the bay a climber will move from the cool surface layer into the warm air above, actually climbing into the inversion. The temperature may suddenly rise 15 degrees within the rise of 20 or 30 feet in elevation.

This warm layer acts as a lid or ceiling above the cool ocean air and its fog. When the stratus comes in, it will rise to this inversion layer and stop, unable to penetrate the invisible barrier of warm air.

As the cool air flows in from the ocean with increasing volume beneath the inversion, it is able, literally, to raise the lid, forcing the warm air layer upward. Thus, the fog roof rises from an elevation of possibly 200 feet to a normal maximum of about 2000 feet. In the early stages of the daily or weekly cycle, the height of the inversion layer can often be estimated by a glance at the Golden Gate Bridge. In the spring the inversion lid may initially be about at the level of the 230-foot-high deck of the bridge, and the stratus flows underneath while the traffic moves through the warm clear air of the inversion layer itself. Over a period ranging from hours to days, the fog roof rises higher until it engulfs even the tops of the 750-foot-high towers, pushing the inversion layer up to that elevation.

High Fog. At about the same time another phenomenon takes place that changes the entire aspect of the stratus on the bay. The ocean breezes and fogs have by now cooled off the layers of air at the surface. The stratus originally was formed when relatively warm air came into contact with the cold water and was cooled to a temperature near that of the water itself. But now the temperature difference between air and water is no longer great enough to create fog.

The fog does not disappear, however. A new fog-forming condition develops. The cooled surface air takes the place of the water as a fog-forming agent, and the fog now forms at the point of contact between the cool surface layer and the warm air above it. The warm air above now meets cold air instead of cold water, but the effect is the same: condensation of the warm air's moisture at the point where it contacts the cold layer.

As a result the fog-forming process is lifted into the air. The bottom of the stratus is no longer in contact with the water but rather with the lower layer of cool air. As the fog roof pushes upward against the inversion, the "fog floor" rises into the sky and itself becomes a ceiling over the Bay Region—a "high fog." The Golden Gate Bridge comes slowly into view again, as if a misty curtain were rising, revealing it from the bottom upward. The ceiling slants upward to the east as the incoming cool air piles against the Berkeley Hills.

As the low surface layer of cool air is replenished from the ocean and increases in depth over the bay and its shores, the temperatures in the region near the Golden Gate continually drop toward the temperature of the water—the low fifties. Thus, San Francisco's coldest spring and summer weather comes not when the stratus is low in the streets, but later when it has been pushed high overhead.

During the summer the inversion, once established, is relatively permanent. It rises and falls with each cycle, but seldom gets as low as in the spring (or in the autumn) when the fog in the Golden Gate rests on the water. When the summer stratus returns after a few clear days, it will usually not form on the water but at some distance overhead, where the layer of ocean air meets the inversion. The typical summer condition is the high fog.

When the stratus is still on the ground (at least in hilly areas), it is possible to distinguish between two varieties: the wet fog and the dry fog. Both types may roll through the streets in almost impenetrable billows. The wet fog (sometimes called "Oregon mist") collects on trees and wires, forming large drops that fall to the ground. Automobile tires swish along the damp pavements and windshield wipers click steadily, barely able to keep the glass clear. A dry fog may be equally dense but does not produce a drip. Even though a driver may scarcely be able to see beyond the hood of his car, no moisture collects on the windshield.

The wetness or dryness of a fog may be the result of various conditions, among them the distance the fog has traveled and the depth of the stratus layer. If the fog-forming process has begun far at sea, the drops of moisture have ample time to grow and merge as the fog-bearing wind moves landward. By the time the fog reaches the coast, it has developed a depth of several hundred feet and the drops have become large enough to produce a steady drip from leaves, wires, and eaves. The drip will be accelerated if the wind is strong, causing the drops to whirl and collide. A dry fog, however, is formed close by, perhaps just offshore or above the coastal hills, and the droplets have not had the opportunity to increase in size.

The wetness or dryness of the fog may be affected also by the temperature. If the temperature is relatively low, the fog will tend to be wet and drippy. If the air is warmer, however, it will have a greater fog-carrying capacity and will not tend to deposit its moisture on wires and windshields.

Because of the inversion lid of warm air above it, the foggy layer usually cannot rise and cool sufficiently to turn into rain, but occasionally a wet fog will be so thick as to deposit a measureable amount of moisture in rain gauges. Even when the fog may not be thick enough or wet enough to deposit moisture in a rain gauge or on open terrain, fog drip beneath trees may amount to rainfall. A stand of trees in a foggy area will not only collect drops from fog blowing directly into the trees, but will tend to stir up the foggy air above them, causing turbulence, as a rocky stream bed stirs up the water above it. The turbulence sets up vertical eddies, and the air eddying down into the treetops deposits moisture on the leaves and branches.

It is the steady drip from this deposit that provides water to nurture the redwood groves in the coastal canyons. Another place of copious fog drips is opposite the Golden Gate in the Berkeley Hills, where the fog deck frequently forms on the upper slopes. Eucalyptus and pine groves planted there long ago intercept large amounts of fog and cause a rainlike deposit of moisture. The fog drip there during the summer months has been measured at an incredible ten inches, an amount nearly half as great as the total annual rainfall in Berkeley.

It has been suggested that planting such groves of trees along barren coastal mountains would greatly increase the annual precipitation and add to the water supply—a possibility of great significance for many areas of water shortage. Careful studies would have to be made, however, to determine whether such plantings might decrease the fog canopy and the fog drip on areas farther inland.

The fog drip is normally the only kind of precipitation that reaches San Francisco in the summertime. Rainstorms approaching the California coast from the Pacific are blocked by the Pacific High, that mountain of cool heavy air centered on the ocean a thousand miles or so offshore. The storms are forced to detour around the northern end of the Pacific High and strike the coast of Oregon and Washington instead. The rare summer shower that strikes the Bay Region usually comes in from the south as an errant mass of warm, moist air moving northwest two thousand miles from the Gulf of Mexico, and perhaps accompanied by thunder and lightning. Summer thundershowers in the Sierra are usually caused by the same kind of moist air from the Gulf, rising and cooling on the undulating western slopes of the range. Occasionally, early or late in the summer, a fragment of a storm moving landward from the ocean may drift through or over the Pacific High and cause light rain in the Bay Region.

Besides contributing to fogs and blocking rainstorms, the Pacific High has other effects on the Bay Region's summer weather. Unknown forces may cause this fickle mass of air to change its shape or strength. Sometimes it sends a long lobe inland over the Pacific Northwest. And since air always blows outward from a high pressure area, strong winds come whipping down from the high plateaus of inner Washington, Oregon, and Idaho across the Cascades and the Sierra to the coastal valleys. Since air warms as it descends (at about 5½ degrees per thousand feet), the temperature of the air masses may rise thirty degrees or more. Hot, dry winds roar down Sierra canyons and out across the Sacramento Valley, blow up bitter dust storms, whistle through the passes of the Coast Range, and flood the Bay Region with warm air.

The ocean fogs are driven back by the hot blasts from the interior. Sunbathers spread out on the lawns of the parks and flock to the ocean shore. For a few days San Francisco has some of its rare beach weather. But this is also fire weather, and the lookouts on Tamalpais and other mountaintops anxiously watch the horizon for signs that the dried-out vegetation has ignited to start the inevitable brush fires.

With the coming of these hot, dry, continental winds

from the northeast, the humidity—the amount of moisture in the air—drops drastically and everything tends to dry out—clothes on the line as well as fruit on the trees and the skin on your face. Wild grasses already dried by the summer sun turn to tinder, ready to blaze at the first spark of a discarded match or cigarette. Even wooden houses, especially those with shingle roofs or walls, dry out and crackle in the heat, and in the cities the fire sirens wail continually.

Sometimes a dry continental air mass moves down into the Bay Region more slowly, causing only a light breeze but still possessing enough power to repel the normal wind off the ocean. San Francisco and other coastal areas enjoy warm, clear weather without the dessicating effects of the northeast wind.

But it was a roaring northeaster that started a grass fire in the Berkeley Hills in September, 1924, and fanned it down the dry slopes to the residential area, consuming hundreds of homes. Most of north Berkeley went up in flames, and the business district was saved only when the northeaster died out at the end of the second day and the cool sea breeze returned, turning back the holocaust. It was another blaze caused by a northeaster that consumed a great part of Mill Valley in July, 1928, before being similarly turned back by the returning sea breeze.

These "fire winds" usually blow themselves out in two or three days (although fire weather has occasionally lasted for weeks), and San Franciscans with chapped lips and sunburned noses welcome back the cool, refreshing vapors from the ocean. On street corners, in downtown elevators, and in neighborhood grocery stores the conversation is the same: "It's good to have the fog back again."

If the Pacific High causes problems when it goes on a rampage, it also throws things out of gear when it does the opposite. Some summers it may weaken and fail to do its normal job of pushing winds toward the coast. San Francisco's fogs normally result from a combination of push and pull (a push from the Pacific High. a pull from the hot Central Valley), but lacking the usual push from behind by the Pacific High, fogs do not have the impetus to go farther than the area immediately inside the Golden Gate. So they settle down in San Francisco, and the city—particularly the oceanside areas—may scarcely see the sun for weeks at a time.

Failing to be cooled off as usual by the ocean wind and fog, the Central Valley remains hot—as do the parts of the Bay Region not under the fog canopy. The fog cycles thus fail to materialize, and while average temperatures in fogbound San Francisco are subnormal, most of the rest of the area will be persistently hotter than usual—all the result of a failure of that vital part of the region's air-conditioning system, the Pacific High.

There are some summers when the San Francisco weather is unusual in the opposite way—scarcely any fog at all comes to the city. A sunny summer in the city is due to a combination of causes, among them a weakening of the Pacific High and subnormal temperatures in the valley, resulting in a slackening of the westerlies that usually blow down the coast in spring and summer. Even during a "normal" summer, occasional hot spells are caused by temporary failures of the Pacific High and the sea breeze, allowing the sun to heat the region to temperatures that would be normal inland.

When the westerlies weaken, they fail to cause the usual upwelling of cold water that creates the fog. Without the upwelling the ocean temperatures are higher than usual—a matter of interest to fishermen as well as swimmers. Many kinds of fish normally found in the cold, upwelled waters off the northern California coast, including salmon and albacore, migrate northward and are not found in the usual numbers offshore. In their place fishermen haul in quantities of fish from the warm southern waters—barracuda, bonito, yellowtail, and even an occasional giant marlin.

Some experts believe that the increased appearance

of sharks off California's beaches is due partly to the warming of the waters in years when the upwelling does not take place. On the other hand, it is possible that hyperactivity by the Pacific High over a period of years (such as occurred from 1942 to 1956), causing increased upwelling and colder water, was responsible for the disappearance of the great schools of sardines that previously swarmed offshore and were the basis of a large fishing industry.

The antics of the Pacific High and the prevailing westerlies affect not only the swimmers and fishermen but even more directly the farmers. The hot northeast winds tend to dry out the fruit on the trees and the vegetables in the fields.

The summer fogs along the coast make the valleys opening on the ocean ideal for growing such cool-weather crops as artichokes and Brussels sprouts. The coastside areas of Santa Cruz and Monterey counties grow most of the artichokes raised in the United States. But when the Pacific High and the prevailing westerlies fail to produce the normal fog, the sun beats down on the coastal valleys and the artichokes tend to dry out, with serious results for growers.

Stock raisers are also affected. Normally the fog drip along coastal areas helps grow grass for grazing cattle and sheep. Without the fog the grass is scanty, and ranchers have to buy more hay to feed their animals. On the other hand, when the fog penetrates too far inland and persists too long, it may spoil grapes in the Napa and Sonoma valleys.

Usually, however, the delicate balance of forces is maintained, giving the Bay Region, at the gateway between the land and the sea, its incomparable combination of continental and maritime climates, dryness and dampness, sun and fog, heat and cold.

At any particular point in the Bay Region the line of demarcation between the two kinds of climate varies with the time of day, the distance from the ocean, and the proximity of "streamlines."

Mill Valley, for example, on the southeast side of Mount Tamalpais, may be damp and drizzly at night and early in the morning but enjoy bright sun in the middle of the day. Yet it is far less foggy there at any time than at Stinson Beach, on the ocean side of the mountain. It has, in other words, more of a continental climate and less of a maritime climate than Stinson Beach.

Go eastward across San Francisco Bay and over the Berkeley Hills to Walnut Creek and the weather becomes steadily warmer and drier—more continental. Continue east over Mount Diablo to Byron and the marine influence is almost entirely gone. The temperature at Stinson Beach may be in the low 50's at a time when Byron broils in the high 90's.

In general the same is true at any point in the Bay Region; a place farther from the ocean will tend to have less of a damp, marine climate and more of a dry continental climate. Yet this principle is sharply modified by the streamlines along which ocean air is channeled through the hills. Two points about the same distance from the ocean may have widely varying climates because of their proximity to passes in the coastal hills.

To take an extreme example, Redwood City is about the same distance from the ocean as Berkeley (actually it is a little closer) but is normally warm and sunny on summer days when Berkeley is fogged in. The reason, of course, is that Berkeley is opposite the lowest gap in ths coastal hills, the Golden Gate, while a high ridge separates Redwood City from the ocean. The damp, ocean air, channeled into a streamline by the Golden Gate, flows directly across San Francisco Bay into Berkeley.

The Golden Gate is the largest and lowest of the gaps in the Coast Range and has the greatest influence on

Bay Region weather, but the other gaps function as "little Golden Gates," funneling ocean weather inland along streamlines (and sometimes allowing land weather to move to the coast, particularly during the hot northeasters). (See frontispiece.)

One of these gaps is not far from Redwood City, causing the summer weather to be cooler than might be expected from its location, although not nearly so cool as exposed Berkeley. Just northwest of Redwood City the San Andreas Fault slices through the hills from the ocean and has created a low-lying area—the Crystal Springs Gap—through which the breeze may penetrate in the afternoon and the fog may flow in the evening. The Crystal Springs Reservoirs, part of San Francisco's water supply, occupy part of this gap. The breeze through the Crystal Springs Gap not only cools Redwood City but extends its refreshing influence as far south as San Jose.

Farther north is a much lower and broader pass between Montara Mountain and San Bruno Mountain, known as the San Bruno Gap, the historic and present route of El Camino Real. The San Bruno Gap is second only to the Golden Gate in its influence on Bay Region climate. The sea breeze and fog pour through this pass to cool off San Francisco Airport and the communities to the south, including San Bruno, Millbrae, Burlingame, and San Mateo.

Immediately north of the Golden Gate there is a narrow gap at Elk Valley, a higher gap above Muir Woods (both of which affect the climate of Mill Valley), and the considerably more important Estero Gap, which funnels winds and fogs from the Bodega Bay area into the Petaluma Valley. The cooling influence of this gap can be clearly felt if you drive north on a summer afternoon along Highway 101. When you round 1600-foot Burdell Mountain and enter the Petaluma Valley, the temperature suddenly drops several degrees and you feel the influence of the sea breeze. The Estero Gap (named for the estuaries at its seaward end) extends its cooling effect as far as Santa Rosa and beyond.

Corresponding to these seven gaps in the western-most range are three in the inner range just east of San Francisco Bay: Niles Canyon and Hayward Pass, which sometimes channel ocean breezes into the Livermore Valley; and Carquinez Strait, the "inner Golden Gate," channeling the streamlines into the Delta and the Central Valley, thus lowering summer temperatures in Stockton and Sacramento.

There are many smaller gaps and passes affecting local areas. The same principle that effects the entire Bay Region holds true within San Francisco, for example; the summer weather is warmer, though modified by streamlines, with increased distance from the ocean. One gap extends eastward from the beach along the line of Geary Boulevard, making that stretch of the Richmond District particularly breezy. Golden Gate Park lies in another gap; its streamline extends inland between Lone Mountain and Buena Vista Peak to the downtown area and blows the hats off pedestrians on Market Street. From the surrounding hills can often be seen a long salient of fog moving slowly eastward through the park.

By far the largest pass through the city, however, is the Alemany Gap immediately north of San Bruno Mountain. The wind and fog often flow from the Lake Merced area along the route of Alemany Boulevard, reaching San Francisco Bay near Hunters Point. One branch of the Alemany Gap extends through Visitacion Valley, channeling the streamlines toward the bay shore and around Bayview Hill to Candlestick Park, where it collides with another streamline from the main Alemany Gap by way of Hunters Point, creating the world's windiest ball park. The convergence of streamlines sets up rapidly shifting eddies and partial vacuums that make a traveling baseball perform in unpredictable ways. Fly balls have been snatched by these vagrant gusts almost out of the mitt of many a frustrated out-fielder.

Oddly, fog will sometimes flow massively through the Alemany Gap to the bay—and Candlestick Park—at times when all other areas around the city, even the Golden Gate, are sharp and clear. The explanation is probably to be found in the varying directions of the wind off the ocean. If the wind comes from a direction to the north of northwest, the long tip of Point Reyes extending into the ocean 35 miles northwest of San Francisco will protect the Golden Gate, and the main force of the wind and fog will bypass that area and strike the San Francisco Peninsula southward near the city line, flooding in through Alemany Gap.

Point Reyes also to some extent protects the beaches at Bolinas and Stinson, making sunbathing there comfortable at times when San Francisco's Ocean Beach is whipped by stiff breezes. The two Marin beaches, particularly Stinson, are also sheltered by Mount Tamalpais. Although Tamalapis is to the leeward rather than the windward, it rises so abruptly behind Stinson Beach to an elevation of some 2000 feet that the beach is usually in a pocket of relatively dead air backed up against the mountain. The winds cannot whip directly across the beach as they do in San Francisco, where the backshore area is low and offers no protection.

Autumn

The only really dependable beach weather comes in early fall, which is San Francisco's real summer. Although elsewhere there is normally a lag of a few days to a few weeks between the summer solstice in June and the period of warmest weather, San Francisco's lag is unmatched in the United States; it amounts to about three months. September and October are the two warmest months of the year.

The reason is to be found in the workings of the same influences that created the coastal fog canopy. After the

Rising fog layer,
north tower of
Golden Gate Bridge

Fog fall and comber,
north end of
Golden Gate Bridge

Fog fall, north side of Golden Gate

Advection layer from above
(photograph: San Francisco Chamber of Commerce)

Advection fog drifting through redwoods (photograph by Gayle Pickwell)

Fingers of summer fog from Golden Gate (official photograph U.S. Navy)

*Fog roof
(photograph by
Gayle Pickwell)*

*Radiation fog
(photograph by
Gayle Pickwell)*

*Cirrus above,
cirrostratus below
(photograph by
Gayle Pickwell)*

*Summer fog
in Golden Gate
(official photograph
U.S. Navy)*

*Cascade pouring
over Muir Woods Gap
into Mill Valley*

*Comber
over Sausalito*

*Comber
rolling into
Sausalito*

*Cascade
over Shelter Cove,
Sausalito*

Fog flowing in from ocean around Mount Davidson; altocumulus and cirrus above

Twin Peaks cascade

"Wave burst," north end of Golden Gate Bridge

Golden Gate from north; fog fall beginning at right

Cumulonimbus

Photographs by
Gayle Pickwell

*Fair-weather
cumulus*

*Cirrocumulus (left)
altocumulus (right)*

Stratocumulus with cirrostratus above

(Photographs by Gayle Pickwell)

Hailstones in grass

Snow on Mount Hamilton

*Lightning, Hayward
(photograph by
Walter Halland)*

*Altocumulus
(photograph by
Arthur C. Smith)*

*Altocumulus
with cirrus
and cirrocumulus above
(photograph by
Edgar A. Smith)*

*Stratocumulus
(photograph by
Edgar A. Smith)*

summer solstice the sun moves steadily south. Its rays do not beat down from straight overhead but begin to slant through the layers of atmosphere, weakening their impact. During the lengthening nights the air cools off; during the shortening days it has not time to warm up. As a result, by September, temperatures in the Central Valley steadily drop from around 100 into the 80's then the 70's. High above the valley in the upper altitudes of the Sierra Nevada, die-hard campers shiver in early-morning frosts.

Meanwhile, momentous changes are occurring over the ocean as well. During the summer the Pacific absorbs the heat of the sun far more slowly than does the land, and it reaches its annual temperature high in early fall. Out over the rolling waters the Pacific High has begun to move south with the sun and ceases to send strong winds racing toward the coast. The Central Valley is no longer hot enough to suck drafts in through the Golden Gate and the other gaps in the Coast Range as it did in midsummer.

When the temperature difference between land and sea decreases, the push-pull effect that created the coastal winds and fogs no longer takes place. As the on-shore winds die down, the upwelling that brought the cold bottom waters to the surface slows to a halt. The great coastal fog bank, which at its maximum reached perhaps 100 miles in width, now fades away to a few wisps along the shore. For a time, as the fog-forming process decelerates, the fogs may come through the Golden Gate low on the water, as they did in the early spring, forming the same fantastic shapes as they roll under the bridge, over islands and shores. Then they cease entirely.

The sun, which during the summer was seen along the coastal areas only intermittently, now beams down in genial autumnal warmth, raising the temperature into the 70's. The pale-faced San Franciscans again flock to the parks and beaches to lie in the sun and relax

[41]

in a pleasant stupor. Surfboard riders paddle out through the warming surf at Ocean Beach to take advantage of the calm surface and the big swells that roll in from the late winter storms of the Southern Hemisphere.

The inversion layer of warm air that hung over the region during the summer—and was lifted at times well above the 1000-foot level by the sea winds blowing in beneath it—now presses lower as the winds die down. Below it the hazy surface layer of moisture-filled ocean air becomes steadily thinner over the land and dwindles to a few hundred feet. On Angel Island the top of 780-foot Mount Caroline S. Livermore may rise out of the marine layer into the clear warm air above, and even the tops of the Golden Gate Bridge towers penetrate the inversion.

Smoke from the chimneys of factories around the the shores of San Francisco Bay rises straight up into the calm windless air, flattens out against the bottom of the inversion layer, thickens as it mixes with the moisture of the sea air, and becomes smog. Gentle afternoon breezes through the Golden Gate push the smog eastward until it is stopped by the Berkeley Hills and collects in eye-irritating volume in the cities along the eastern shore. If the afternoon breezes do not materialize or are countered by sluggish movements of continental air from inland, San Francisco itself is wrapped in gray, noxious vapors and visibility may drop to a few hundred yards. At times the smog over the bay has been so thick as to endanger shipping and cause the fog horns to be set off.

Other unusual atmospheric effects are caused by the low inversions of fall. At times the breeze through the Golden Gate may falter, then recommence, bringing thin layers of cool sea air in beneath the hazier, warmer air around the bay. The result is to lift the haze a short distance off the water. This process may occur a number of times in a few hours until there are several layers of

[42]

alternately hazy and clear air, all clearly visible against the hills around the shore.

The bottom of an inversion layer sometimes has reflecting qualities and may cause mirages. An apparent vertical elongation of objects on the horizon is a "superior mirage," an effect also known as "looming." A ship or a cloud or the Farallon Islands may appear to be several times as tall as they actually are. As seen from such places as Stinson Beach, ships passing offshore can be reflected upside down, causing a double image. On one occasion, from a boat in Richardson Bay off Sausalito, I saw an "inferior mirage" (in which the reflection appears below the object); the tall buildings of downtown Oakland appeared in the water north of the Bay Bridge.

As the sun continues southward, the nights grow longer and the days shorter and the atmosphere gradually cools, bringing bright fall colors to the elms and maples on the creeks of Tamalpais and a hundred other streams from Ben Lomond to the Russian River. The air is full of the spicy aromas of autumn and the smell of burning leaves. In the vineyards of Napa and Sonoma the grape harvest is over and the leaves of the vines turn bright reds and yellows and purples.

But the weather is still unpredictable. The Big Game in late November may be held under a hot sun in a stadium that resembles an oven—or in brisk, snappy, fall air appropriate for overcoats and the resounding thump of shoe leather on pigskin.

Winter

The first storms to strike the Bay Region may come as early as September or October—showery masses of clouds that escape through the weakening Pacific High or move north from the offshore areas of Mexico, occasionally with thunder and lightning displays. Often, however, even after the first rains, Indian summer lasts into November, which in coastal areas is usually warmer

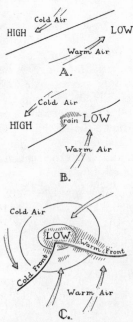

than breezy April. Sooner or later, however, the Pacific High will have moved far enough south to open the way for transoceanic storms.

The storms are born out of the conjunction of warm and cold masses of air far out on the rolling surface of the ocean. A thousand miles off the China coast a mass of cold air from the polar regions may move south over the Pacific, encountering a mass of warm southern air. The two masses do not immediately merge but may flow past each other in opposite directions. At some point a bulge of light, warm air may rise over the edge of the heavy, cool air. As it rises it creates a low-pressure area, which sucks in wind from all directions to replace the air that has risen.

The inrushing air, affected by the Coriolis force, cannot move straight into the low pressure area but curves to the right, causing the entire air mass to begin to move around in a counterclockwise direction, forming an eddy. This counterclockwise circling of air masses around a low-pressure storm center is called a "cyclone," although its winds are normally much slower than the tropical cyclone or hurricane. At the center of the cyclone the rising warm air cools and its moisture condenses into rain. A storm is born.

The entire revolving mass of air moves with the prevailing westerlies toward the Pacific coast of North America, continuing to draw fresh supplies of air into it.

[44]

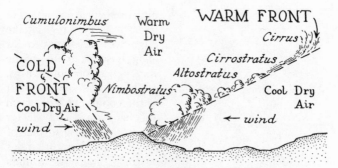

If the cyclone at this stage of development were to pass over the Bay Region, the cool air already over the area would be succeeded by the warm air near the center of the cyclone, followed by another mass of cold air. Sometimes these air masses are visible as they pass overhead. The advance edge of each of these masses is known as a "front." Thus the cool air in the Bay Region would be succeeded by a warm front, then a cold front. The entire process is sometimes called a "frontal system."

The warm air mass near the center of the cyclone, being lighter, tends to slide up over the edge of the cold air ahead of it. As it rises it cools off and its moisture condenses, creating a cloud. The form taken by the cloud depends on the height at which it condenses. The advance edge of the warm air mass, riding over the cool air, will attain heights of perhaps 20,000 to 30,000 feet. At this great elevation the temperature is so low that the moisture condenses into fine particles of ice, forming a cirrus cloud—the herald of a storm—a thin, hazy sheet of ice vapor barely visible high in the sky, perhaps forming a ring around the sun or moon.

As the storm moves on, the clouds lower and thicken to altostratus, then pass through regular stages to alto-

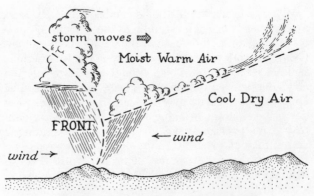

storm moves ➡

Moist Warm Air

Cool Dry Air

FRONT

← wind

wind →

OCCLUDED FRONT

cumulus, cumulus, stratocumulus, and finally the low, dark nimbostratus or rain cloud. These later stages are usually not distinguishable by observers standing directly beneath them, but appear only as lowering, thickening clouds.

As the warm front at the center of the cyclone passes overhead, the air will be moist and full of mist—"hazy sunshine." Then comes the trailing cold front, marked first by towering cumulonimbus clouds and showers, where its impact causes the warm air to condense, then by clear cool weather as the clouds move on to the east.

Often, however, this "normal" sequence does not take place in the Bay Region. If the storm was born far out on the Pacific, as most of them are, it reaches the coast at a later stage of maturity. The trailing cold front tends to crowd in on the warm air mass at the center, scooping beneath it to join the cold air mass ahead. Thus the warm air mass is "occluded" or closed off. It no longer reaches the ground but still drops its precipitation through the cooler air from high overhead. The "textbook storm" sequence outlined above is abridged by the elimination of the warm front, and the nimbostratus is succeeded directly by cumulonimbus.

Whether or not the textbook cloud sequence appears at all depends on such variables as the amount of moisture in the storm, the depth of the storm, and the circulation of high-level winds. A shallow storm, for example, will not reach altitudes high enough to form cirrus and may be preceded only by the lower altocumulus. The appearance of cirrus may indicate the approach of a deep, moist storm bringing heavy rain. Cirrus clouds may pass overhead, however, without heralding a storm; they may mark the edges of a storm passing far to the north or the remains of an old storm that has lost its energy at sea and is dying.

Often the coast will be hit by a series or family of storms, one behind the other, in such close succession that the weather does not clear between them. During the interval the sky may be covered with various types of clouds from low stratus to altocumulus.

Even though in the Bay Region cirrus is not a good guide to coming storms, there are other indications that are more reliable. The principal herald of a storm is a change in wind direction accompanied by low cloudiness. On the skyscrapers of San Francisco the flags, which usually are stiffly whipped eastward by the west winds, first droop listlessly on their masts, then begin to billow out toward the north. Rising smoke from factory chimneys similarly drifts northward.

The reason for the wind shift is the counterclockwise circulation of air around the low-pressure center of an approaching storm. With the coming of a low from the ocean, the winds circling the low will come to the Bay Region from a southerly direction. The exact direction of the wind depends on whether the storm center is to the northwest or southwest. If the storm center is approaching the coast of northern California or Oregon, the winds will come at first from the southwest. If the storm comes from the west or southwest, the first winds will come from the southeast.

As the storm moves on to the east, the direction of

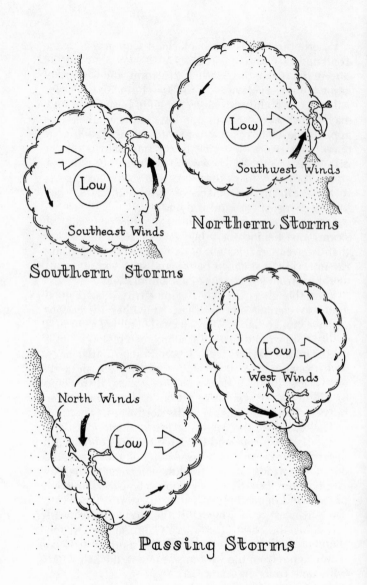

Southeast Winds

Southern Storms

Low ⇨

Southwest Winds

Northern Storms

North Winds

Low ⇨

Low ⇨

West Winds

Passing Storms

the wind changes to correspond with the stage of the storm. Winds from a southerly direction indicate that the storm has yet to reach its peak. Winds from a northerly or westerly direction indicates that the storm is on its way out.

I lived for some years in a house at the top of Telegraph Hill in San Francisco, where the course of the storm was easily visible; raindrops on the south windows meant a storm in its early stages. But rain striking the north or west windows was an indication that the storm was passing away. The same test cannot accurately be applied to houses in lower areas, however, as hills or nearby buildings may cause wind eddies.

Similarly, it is sometimes possible to predict the weather by sniffing the wind. At the Telegraph Hill house, for example, a pungent aroma from the coffee roaster near the west end of the Bay Bridge was an indication of a movement of air from the south—and a possible storm. Sometimes the wind would come from a different southerly quarter, bringing the soapy smells of the copra docks at the southern end of the water-front. A salty smell meant an ocean breeze from the west through the Golden Gate, bringing fog or the damp air off the ocean. The acrid scent of factory smoke indicated a movement of air from the industrial areas of the East Bay, from Emeryville to the refineries of Richmond and beyond. If the movement of air was rapid, it would bring clear, dry continental air from the high regions of the Great Basin—Nevada, Idaho, inner Oregon. If the movement was sluggish, it would sometimes be sufficient merely to counter the normal ocean breeze and carry the industrial smoke into the city, bringing smog.

The same conditions that cause a vast variation of summer climates within the Bay Region also create equivalent variations in winter climates. Just as certain communities get immense amounts of summer fog while nearby points are usually clear, so certain spots get far more rain than others.

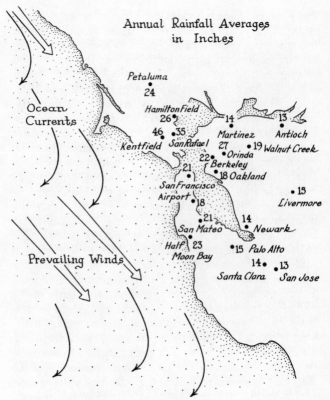

Annual Rainfall Averages
in Inches

Ocean
Currents

Prevailing Winds

Petaluma
24

Hamilton Field
26

46 35
Kentfield San Rafael

14   13
Martinez   Antioch
27 19 Walnut Creek
Orinda
22 Berkeley
18 Oakland

21
San Francisco
Airport 18

21
San Mateo

Half 23
Moon Bay

15

14 Newark

15 Palo Alto

14 13
Santa Clara San Jose

15
Livermore

San Francisco, for example, gets an average of 21 inches of rain annually, but Kentfield, nine miles north of the Golden Gate, is pelted with more than double that amount—46 inches. The Bay Region weather station with the smallest annual rainfall figure—arid San Jose, with 13 inches—is only 20 miles from the point with the heaviest rainfall—Boulder Creek, which is annually deluged with 60 inches.

The rainfall pattern is more complicated than the fog pattern. The summer fog comes from the ocean and dissipates to the east, but the rain comes from several directions during the course of a single storm. The most prolonged rains often form a family of storms following one another in rapid succession to strike the area from a southerly direction. As the air masses rush in from the south, they are funneled in through the valley of the San Lorenzo River, north of Santa Cruz. Trapped in the river canyon and forced to rise with the terrain, the winds cool as they rise, and their moisture condenses, drenching communities in the valley—including Boulder Creek. A similar phenomenon takes place in Marin County when the southerly winds collide with 2600-foot Mount Tamalpais and are forced to rise, dropping heavy rains on the south side of the mountain. If rain measurements were taken in exposed positions on the south slopes of Tamalpais, the totals would probably be even greater than in Kentfield, which is around the mountain to the east on the edge of the heavy rainfall zone.

Dry San Jose, in the center of the Santa Clara Valley, bounded by mountains on the south, east, and west, is in the "rain shadow" of these mountains—protected from the rain-bearing winds. Even Palo Alto, at the northern end of the valley, receives only an inch more than San Jose. Thus, in general, the southern slopes of mountains receive more rain than the northerly slopes. A valley open to the south (such as the San Lorenzo) will be deluged, while a valley opening to the north (such as the Santa Clara) will be relatively arid.

In the later stages of the storm, as the wind shifts around to the west, the rain will come from that direction, pouring down on the western slopes and leaving the eastern slopes relatively dry. Thus Half Moon Bay, on the seaward side of the mountainous San Francisco Peninsula, receives more rain than San Mateo on the east side of the same mountains.

Farther inland, Walnut Creek, partly sheltered from southeast winds by 3849-foot Mount Diablo, receives only 19 inches of rain as against 27 inches for neighboring Orinda, situated in a valley partly open to the south.

[51]

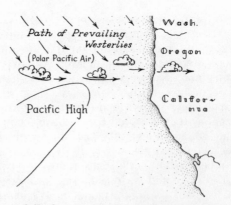

The rainfall of a community is affected not only by its position in relation to nearby hills or mountains but also by its latitude. Because of the Pacific High, which often blocks approaching weather almost as effectively as a mountain range, storms are more likely to hit the coast north of San Francisco than south. Storms moving around the north end of the Pacific High may hit the coast of Oregon, and often only their southern fringes will strike the Bay Region. On occasion the Weather Bureau has been able to specify "rain north of the Golden Gate." As a result, Hamilton Air Force Base, on the northwest shore of the bay in Marin County, gets nearly one-fourth more rain than San Mateo, also on the shore of the bay some 34 miles south.

The rainfall pattern, like the summer fog pattern, is also influenced by gaps in the coastal hills. The rain-bearing westerly winds in the late stages of a storm sweep through the Golden Gate and strike Berkeley with one-fourth more rain than Oakland, which is partly sheltered from ocean winds by the hills of San Francisco. Similarly, the other gaps will funnel rain-bearing winds through the mountains and cause greater rain in points along their path to the leeward. North-south gaps will have a similar effect during the early stages of a storm.

Just as the Bay Region's rugged hill-and-valley contours cause wide variation of rainfall throughout the region, so similar contours on a smaller scale cause variations within a community itself, and the report of the community's weather station does not necessarily represent average conditions in the community.

Berkeley's rain gauge is on the University of California campus at the foot of the Berkeley Hills at an elevation of several hundred feet, where the air has been forced to rise and deposit more of its moisture than fell on the lower parts of the city. In San Francisco the weather reports that appear in the newspapers are based on observations made in only one place; the roof of the Federal Office Building at the Civic Center (although long-range data are also collected in Golden Gate Park). But the city's hills and valleys doubtless cause variations similar to those that take place on a larger scale in the Bay Region as a whole. More rain can be expected on the southern and western slopes of Mount Davidson and Twin Peaks—and perhaps even Nob and Telegraph hills—than on the northern slopes; more in the oceanside Richmond and Sunset districts than in the hill-sheltered areas east of Twin Peaks. The Mission district, in the rain shadow of the hills partly surrounding it, can be expected to be drier than less-sheltered districts such as the Western Addition or Pacific Heights.

Unfortunately, however, these variations and similar ones in other parts of the Bay Region have not yet been measured. Consequently, much is still unknown about the Bay Region's unique weather patterns.

The types and amounts of rainfall in the Bay Region are strongly influenced by the direction from which storms approach. A storm center may move toward the coast from any part of the Pacific between the Gulf of Alaska on the north and the Hawaiian Islands on the south. Storms from the Alaska region begin with a southwest wind and tend to bring cold air. They occa-

[53]

sionally bring snow to Bay Region mountaintops, from a few inches on Mounts Diablo and Tamalpais to a sugary sprinkling on Twin Peaks. The rare snowflakes that fall in downtown San Francisco usually melt as they hit the ground, but one frigid February day in 1887 residents were delighted when nearly four inches covered the rooftops—a record that has not yet been surpassed.

The Alaskan storms bring cheer to the hearts of Bay Region winter sports enthusiasts, because these may leave snow as low as the 3000-foot elevation on the Sierra Nevada, and at higher altitudes usually deposit layers of dry powder snow, perfect for skiing. Hawaiian storms, on the other hand, usually begin with a south-east wind and are accompanied by warmer air. Because of the greater capacity of these warm air masses to hold water, they dump heavier rains on the lowlands and mountains as well. The Sierra snow line will thus be far higher—perhaps 8000-10,000 feet.

The warm Hawaiian storm is the despair of skiers, for its copious rains turn the ski runs to slush, and even at the highest elevations it leaves soft snow into which skis may sink to a depth of several feet. It was a soft-snow, Hawaiian storm that buried and isolated Southern Pacific's de luxe passenger train, the "City of San Francisco," in Donner Pass for three days in January, 1952.

Hawaiian storms, because of both their greater water content and their ability to melt Sierra snows, bring the greatest flood danger. It was a series of such storms that arrived just before Christmas, 1955, swelling the rivers to overflowing, inundating Yuba City, and deluging California with the greatest floods on record. For months afterward the Sacramento and San Joaquin rivers spread a coating of muddy brown flood waters across the surface of San Francisco Bay.

In late fall or early winter—after the first rains have soaked the soil, the storms have departed, and the air is clear, cold, and still—conditions are right for another change of mood in the atmosphere.

Just as the patterns of summer fog and winter rain at any particular point in the region are determined by its location in relation to the ocean and nearby mountains, so winter temperatures are strongly influenced by the same elements of location. The moderating influence of the Pacific is the greatest single conditioner of Bay Region temperatures. The ocean varies in temperature only a few degrees from winter to summer; consequently it is cooler than the land in the summer and warmer than the land in the winter.

During the frigid winter weather the land temperatures drop far below those of the ocean. Winds from the west bring warmth from the ocean to the land, and the winter temperatures of the Bay Region as a whole are high above those of inland regions far from the ocean's direct influence. Similarly, the coastal part of the Bay Region is the warmest section of the region in the wintertime, as it is the coolest in summer. The coastal warmth is prevented from moving inland by successive ranges of hills, and each valley farther inland is colder. Thus, Half Moon Bay in the winter will be warmer than San Mateo, which in turn is warm compared to Walnut Creek. Coldest of all are the localities east of the Coast Range. Like the patterns of fog and rain, this winter temperature picture is modified by the passes in the hills and mountains, which tend to bring warmer coastal temperatures inland.

As the nights lengthen, the earth and the air quickly lose their heat after sundown, radiating it outward. The coldest air accumulates in the lower areas farthest from the ocean. After the first winter rains, the cold air in these places absorbs moisture from the damp earth. In the frigid hours before dawn the temperature sometimes drops so low that the moisture condenses into vapor, forming radiation fogs.

Because the radiation fogs form most readily in low damp places, such as the Delta, where tules and other marsh plants grow, it is commonly known as "tule fog." ("Tule" is an Aztec-Spanish word for "bullrushes," and the "e" is pronounced.) Often the tule fog forms only a thin layer a few feet deep in the lowest areas. Sometimes the backs of grazing cattle can be seen eerily rising out of the mists. Pedestrians may walk down the road nearly blinded by thick vapors while the sun is shining at treetop level.

During a particularly cold, windless spell the tule fog may accumulate night after night until it is several hundred feet deep in some areas, particularly east of the Berkeley Hills. Because air tends to move from a cold to a warm area (from high pressure to low), the fog-bearing, inland air begins to drift toward the warmer coastal regions. If the cold, windless weather continues for days or weeks, the thick white vapors pile up in deep drifts east of the Berkeley Hills, pour through Carquinez Strait and other gaps in the hills, move across the bay to San Francisco and Marin, and roll outward through the Golden Gate, thus flowing in the opposite direction from the summer fogs, although at a much slower rate. At times the tule fog may form over and around the bay itself and then move slowly oceanward.

A rough compensation takes place: inland areas that are usually fog-free in the summertime may in winter be deep in tule fog at a time when the coastal areas bask in clear sunshine. Within San Francisco, dark inland vapors may bring gray gloom to Market Street while out beyond the city's central spine of hills the oceanside Richmond and Sunset districts enjoy the sun.

As with the summer fogs and the winter rains, the gaps in the hills determine the local weather in detail. The Richmond District, nearest the Golden Gate, may be hit by a tule fog spilling over from the Gate, while the Sunset is clear—unless the vapors are thick enough to roll westward through Alemany Gap to the ocean, spreading over into the Sunset.

Sometimes the radiation fog reaches such depth and density that San Francisco and Oakland airports are socked in and the only clear airfield in the region may be at coastside Half Moon Bay; highway accidents always reach a peak during a tule fog; ships probe gingerly through the murk. At such times San Francisco Bay resounds with an immense orchestration of fog horns, bells, sirens, and ship whistles. Most shipping accidents, including the disastrous sinking of the liner *Rio de Janeiro* in 1901, with the loss of 130 lives, have occurred during thick tule fogs.

If the fog burns off under the heat of the morning sun, San Francisco becomes visible from the top downward. First appear the upper slopes of Twin Peaks, Mount Davidson, and other summits, rising like islands above a sea of white vapor. Then the tops of the hotels on Nob Hill and the skyscrapers of the financial district emerge from the mists; finally the fog lingers only over the bay, where the funnels and superstructures of ships may appear long before the hulls of the vessels are visible.

Usually the tule fogs last for only a few days, then are driven away by the resumption of the normal ocean breeze, perhaps followed by another storm.

Just as in the summer, spring, and fall the normal weather-making influences are upset when continental air masses move down to the Bay Region from the northeast, bringing hot dry winds and clear skies, so a similar phenomenon during the winter will abruptly replace normal weather conditions for a few days.

Over large parts of the globe, great rivers of air move in northerly and southerly directions, as cold, polar air masses flow toward the Equator to replace the rising, warm air of the tropics.

Over the continental United States during the winter there are usually two or three rivers of air moving north and south. Often the largest of these is a current of cold, Arctic air moving down from Canada. But the flow of cold air from the north must be balanced by a return flow of warm air from the south. Thus, one part of the country may swelter while another part freezes.

On occasion southern air moving north may bring

summerlike warmth to California in December, while east of the Rockies a great current of icy polar air flows down from Canada, bringing sharp, blue skies and sub-zero temperatures to many parts of the Great Plains. Thus, winter temperatures in the Bay Region may rise

RIVERS OF AIR

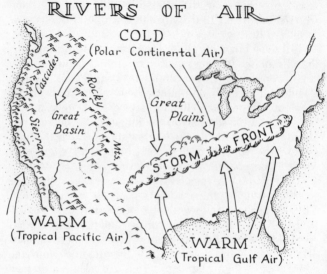

COLD
(Polar Continental Air)

Cascades

Rocky Mts.

Sierra Nevada

Great Basin

Great Plains

STORM FRONT

WARM
(Tropical Pacific Air)

WARM
(Tropical Gulf Air)

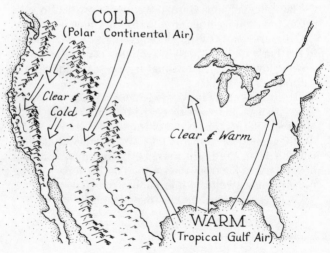

COLD
(Polar Continental Air)

Clear & Cold

Clear & Warm

WARM
(Tropical Gulf Air)

into the 70's while the residents of South Dakota are having trouble chipping the ice off their windshields at 20 below zero.

Along the east coast may come a third great current—warm, moist air moving up from the Gulf of Mexico. When this warm current encounters the cold air from the north, the result is usually spectacular—the birth of cyclonic storms bringing wind, rain, and snow across an arc of 2000 miles from Albuquerque to Denver to New York to Boston.

At times these north-south currents overflow their channels and spread perhaps a thousand miles in either direction. Sometimes the icy Arctic air flowing down from Canada to the Missouri Valley may swing west to the corridor between the Rockies and the Sierra. Then the Great Basin states—Utah, Idaho, Nevada—will have frigid weather while the Midwest basks in warmer air moving up from the south.

More rarely, the polar current of air will send icy tongues down west of the Sierra into the Central Valley and over the Coast Range, stirring up great clouds of dust in the valley farm lands, sending the temperatures down into the 30's, scuffing up whitecaps and frigid spray on San Francisco Bay, howling up Market Street, and sending shoppers and commuters scurrying for the warmth of stores and offices. Having been diverted west, the Arctic stream of air no longer blasts the Great Plains, and newspaper readers in that region, enjoying unseasonal warmth, gloat over the cold spell in California.

Usually two or three times a winter there comes such an invasion of polar air in the Bay Region. This brilliant, dry air brings the clearest weather of the year. On other occasions, the polar air mass does not blow into the Bay Region with great force, but seeps slowly across the Sierra, bringing low temperatures, frost, and ice to the region without clearing away the haze and fog. Usually after a few days the river of cold air returns

[59]

to its normal channel and is replaced by warmer marine air; the Bay Region's usual winter weather returns.

The most violent atmospheric force to occur in the Bay Region—or any place on earth—is the tornado. Although Bay Region tornadoes are midgets compared to the giant twisters of the Middle West, they have nevertheless been powerful enough to cause serious damage. Like most other kinds of storms, tornadoes are caused by the impact of different kinds of air masses. Under certain circumstances a cold, dry air layer moves in above a damp, warm one; the warm air, struggling to rise, upends both layers, and the resulting eddy develops into a tornado—a giant funnel-shaped whirlwind that sucks up everything in its path.

The greatest tornado belt on the globe is in the Mississippi Valley, where damp marine air moving north from the warm Gulf of Mexico meets cool, dry air coming over the Rockies or down from Canada. The same effect takes place on a much smaller scale in California, when cold, dry air moving in from the north or west on the heels of a rainstorm hits the warm moist air of the storm center, creating an eddy that gives birth to a tornado.

Rain storms do not necessarily precede tornadoes, however. The big winds can occur when the air is calm and oppressively humid and a cold air mass strikes it at an angle to form the necessary eddy.

The tornado as a whole moves steadily across the land at a rate of from 5 to 60 miles per hour, but the speed of the whirling wind near the ground has been estimated at an incredible 500 miles per hour. The tornado, as distinct from other kinds of winds, wreaks its devastation in a narrow path where the tip of its funnel touches the ground.

Several dozen tornadoes have been recorded in California at the rate of about one a year. The most disastrous was one that struck Sunnyvale on January 11,

1951, and cut an erratic swath of destruction from 100 feet to four blocks wide.

As spring advances into summer the hot inner valleys are the scene of another kind of whirlwind far more common in California than the tornado—the dust devil. Under a blazing summer sun on a windless day a patch of barren ground gets much hotter than does ground covered by foliage. Over the superheated spot there forms a "bubble" of hot air, something like the bubble on the bottom of a pan of water on a stove.

The "bubble" finally gets large enough to burst loose and rise, leaving a partial vacuum below. Cooler air currents rushing in to fill the vacuum collide with such impact that they whirl into a twisting tornado shape, picking up dust, papers, leaves, sand, and sometimes small buildings. Fortunately, however, the dust devil lacks the tornado's devastating power.

The extreme heat that causes the dust devil does not normally last for many days at a time. The valley heat draws ocean wind and fog through the Golden Gate and the other gaps in the low coastal hills, setting off the region's natural air-conditioning system. The ocean breezes gradually diminish the valley heat, while the fog hangs thick over San Francisco Bay, setting off the great chorus of fog horns. After a few days of breeze, the valleys are cool again, the great winds are no longer sucked in from the coast, and San Francisco Bay once again scintillates in the summer sun.

The earth has made one more circuit around its blazing star; the rivers of air flow across the face of the planet in their summer courses; and here, where the Pacific has breached the mountains along the edge of the North American continent, the aerial forces of the land and the sea collide in a visible war of the elements, ranging over mountains and valleys, through towns and cities, and across the face of the moving waters.

CLOUDS

There are three large families of clouds: cumulus, cirrus, and stratus.

Cumulus clouds are "accumulations" or bunches of condensed water vapor, flat on the bottom and billowy above. They are formed as a mass of heated air rises and cools and its vapor condenses into visible form. Each cumulus cloud can be regarded as the top of a column of rising air. The flat bottom indicates the altitude at which the rising vapor has cooled to the point of condensation.

Low cumulus clouds at altitudes of a few thousand feet take on individual shapes, drift across the sky with the breeze, usually from the west, are signs of fair weather, and are referred to as "fair-weather cumulus." Altocumulus, or high cumulus, are similar clouds at altitudes above 6000 feet, forming a mottled pattern in the sky. Cumulus clouds broken up by a high wind into wisps are called fractocumulus.

Cumulus may grow in size by boiling upward into rising domes and battlements until they become "thunderheads," although there may be no thunder at this stage. These towering cumulus clouds sometimes form around the shores of San Francisco Bay during late winter and early spring, rising above the surrounding mountains to reflect and refract the sunlight in resplendent architectural structures that slowly change in shape.

If these cumulus structures continue to grow upward, their rising tops reach the upper regions at perhaps 25,000 feet or more, where they flatten out into anvil shapes. This is the true cumulonimbus, the most spectacular of all clouds. "Nimbus" means "rain," and the air boiling upward many thousands of feet through these massive structures is cooled until the drops of moisture become large enough to fall in the form of rain, frequently accompanied by thunder and lightning.

Cumulonimbus clouds are rare in the Bay Region, but are most likely to occur in the early fall when the westerlies are weak and warm masses of humid air move northwest from the Gulf of Mexico.

Cirrus clouds, which constitute the second principal family, are always in the high altitudes, ranging to 30,000 feet and more. They are so thin as to allow sunlight through, casting no shadows. They are composed of ice crystals formed in upper areas where the air is below freezing. Cirrus means "curl," and often these clouds appear as wisps or curls sometimes called "mare's tails." If they are bunched together in clusters or ripples they are cirrocumulus, forming a "mackerel sky" or "buttermilk sky." Cirrus clouds indicate a high layer of air with a different origin and greater moisture content than lower layers. They frequently—but not always—precede a cyclonic storm (see pp. 44-47).

The members of the third family—stratus clouds—form in layers at any altitude. This is the only kind of cloud that normally appears in the Bay Region in the summer. Stratus at ground level is fog. Lifted off the ground, it becomes high fog. From 6000 to 17,000 feet in the air it is altostratus, and the thin sheets of cloud that form above that height are cirrostratus. Stratus forms in a layer of air that has cooled uniformly (rather than in bunches or accumulations) either by rising or by contacting a cold surface, such as an ocean current or cold ground. A slight "bunching" of stratus clouds makes them a stratocumulus. If various types of clouds merge and lower, the resulting dark mass is nimbostratus, the ordinary raincloud, usually responsible for several hours of rain—in contrast with the showers of shorter duration caused by cumulonimbus.

There are numerous other combinations and subtypes of these various kinds of clouds. The U. S. Weather Bureau publishes a cloud chart with 36 illustrations of various cloud types. By sending ten cents for this chart to the Superintendent of Documents, U. S. Government

Printing Office, Washington 25, D.C., you can learn to recognize the common kinds of clouds as well as such specialized forms as "cumulonimbus capillatus" and "altocumulus translucidus undullatus."

ACTIVITIES

The uniquely varied weather of the Bay Region provides unparalleled opportunities for students and amateur observers to make valuable contributions to knowledge of the local weather and climate. The U. S. Weather Bureau is concerned with large-scale weather, and cannot maintain stations to observe each of the region's microclimates—of which there are dozens within San Francisco itself. The downtown Weather Bureau station in San Francisco, for example, cannot record fog conditions throughout the city. Systematic observations by amateurs in these areas would add significantly to an unexplored field of knowledge.

Particularly valuable studies can be made af summer fog cycles, correlating the fog conditions at various points in the Bay Region with temperatures at coastal and Central Valley points.

Another possible area of study is the rainfall pattern in various microclimates. How much more rain is there on one side of Twin Peaks, for example, than on the other side?

Valuable aids in weather observation and prediction (obtainable at most scientific supply houses listed in the yellow sections of telephone directories) are the thermometer, the barometer, the rain gauge, and the anemometer.

The barometer measures changes in air pressure. The most common type is a mercury barometer—a tube about a yard long, sealed over at the top, with the bottom in a pan of mercury. If a vacuum is maintained in the tube, the outside air pushing down on the mercury

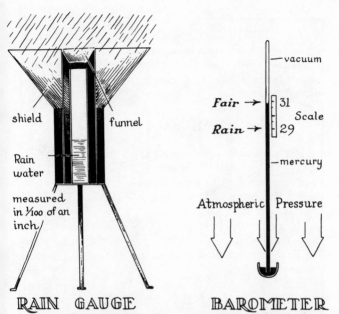

RAIN GAUGE

BAROMETER

in the pan will force the mercury in the tube up to a height of about 30 inches at sea level.

As the air pressure changes, the mercury will fluctuate up and down a few tenths of an inch. Cold, heavy air, for example, will press down with greater force on the mercury in the pan than warm, light air, thus sending the mercury in the tube up. The approach of cold weather is therefore characterized by a rising barometer. As a cyclone-type storm approaches with its lighter air, the barometer will fall.

The aneroid barometer registers changes in air pressure on a vacuum box. As the box top is pushed down by high pressure, the result is indicated on a dial.

Rain gauges are of many types, but the most accurate are those with a large funnel and a narrow receptacle. The height of the water in the receptacle must be divided by the proportion of the funnel's area to that of the receptacle. Thus a funnel four times the area of the

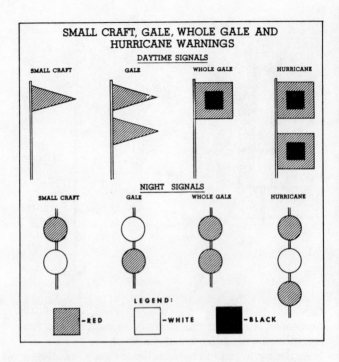

SMALL CRAFT, GALE, WHOLE GALE AND
HURRICANE WARNINGS

DAYTIME SIGNALS

SMALL CRAFT   GALE   WHOLE GALE   HURRICANE

NIGHT SIGNALS

SMALL CRAFT   GALE   WHOLE GALE   HURRICANE

LEGEND:

—RED   —WHITE   —BLACK

receptacle would require that the height of rain water be divided by four to get an accurate measurement.

The anemometer measures wind speed by a system of cups that rotate horizontally, indicating the speed on a dial.

Storm warnings are posted in San Francisco on top of the 26-story Telephone Building on New Montgomery Street near Howard. Small-craft and gale warning flags are also flown at the Marine Exchange lookout station on the end of Pier 45 near Fisherman's Wharf. They reflect indications of expected high wind, primarily for the benefit of navigators of ships and small craft.

In most large cities the U. S. Weather Bureau maintains stations that can be visited by appointment. In San Francisco the State Climatological Office is in the Fed-

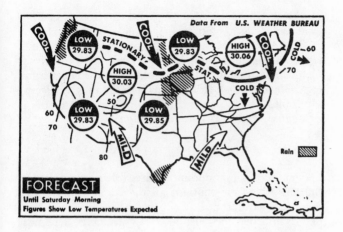

FORECAST
Until Saturday Morning
Figures Show Low Temperatures Expected

eral Office Building at the Civic Center; the Forecast Center at San Francisco Airports coördinates weather information from many areas. There is also a station at Oakland Airport. Several hundred small weather stations are operated throughout the state, where volunteer observers collect daily records from barometers, thermometers, rain gauges, and anemometers.

The official U. S. weather map is published every day by the Weather Bureau in Washington, giving weather conditions throughout the country. A simplified form of the weather map, containing a forecast, appears in many newspapers. The weather map shows conditions for July 28, 1961. Low- and high-pressure areas are indicated in circles, and the number in each circle indicates the height of the mercury in the barometer tube.

Thin lines on the map connect points of equal temperature. The thick lines connect points of equal barometric pressure and are called isobars. The edge of a high or low pressure system is a front. The general direction of winds can be presumed by recollecting that winds blow into a low pressure area, circling counterclockwise, and out of a high pressure area, circling clockwise. The cold front moving down from New England can be expected to bring cooler weather to the Middle Atlantic states. A low pressure trough of heated, rising air extends from Mexico northwestward across California, drawing coastal winds inland and creating fog along the coast. Not shown on this map is the Pacific High offshore. When this or any high pressure area is elongated, it is referred to as a ridge.

SOME DEFINITIONS

Humidity is the amount of invisible water vapor in the air. When air contains all the water vapor it can hold, it is saturated. Beyond that point the vapor condenses and becomes visible as fog or clouds. The most common way to measure water vapor is in terms of relative humidity, which is expressed as the percentage of saturation. If air holds half the moisture it is capable of holding, the relative humidity is 50 per cent. If air is saturated, the relative humidity is 100 per cent.

Because warm air is capable of holding more moisture than cold air, as the air cools, its relative humidity increases.

Because of frequent ocean winds and fogs, the average relative humidity of the coastal section is high, except when the dry northeasterly winds bringing fire weather may send the humidity down to 20 per cent in San Francisco. The record low is 15 per cent.

Dew point is the temperature at which condensation occurs when air cools and forms dew clouds or fog.

When the temperature is low enough to freeze the vapor as it condenses, **frost** is formed. In freezing weather the dew point is the temperature at which frost will form; this is a vital matter to farmers whose crops may be damaged.

Frost is rare in the coastal areas because of the moderating influence of the ocean; it is more common in areas away from the ocean.

Rain falls when the water particles in the saturated air of a cloud increase until they cluster together to form drops. When the water vapor condenses at temperatures below freezing, the result is **snow**. If rain or slushy snow falls into a layer of air that is below 32 degrees, the drops will freeze into **sleet**. When raindrops hit an updraft in a thundercloud and are carried high into freezing air above, they congeal into **hail**.

Lightning occurs when the turbulent air currents in a thunder cloud cause a collection of positive electrical charges in one part of the cloud and negative charges in another. The attraction of the charges becomes so great that they rush together in a massive charge that shoots from cloud to cloud—or cloud to ground. **Thunder** is the sound caused by the sudden expansion of the air heated by the lightning.

IS THE CLIMATE CHANGING?

Talk that the climate is changing is inevitably heard whenever there is unusual weather for a prolonged period—whether it be heat or cold, rain or drought. In recent years it has been popular to attribute apparent changes to nuclear testing but the largest man-made explosions are puny compared with the great climate-making forces; they are unable to affect the weather except, conceivably, in the immediate area at the time of the detonation.

There are other influences, however, operating to affect the climate. It is probable that in the San Francisco Bay Region within the past century—as in other built-up regions—there has been and will continue to be some raising of the average temperatures because of urban influences. Smoke, steam, reflection, and radiation from heated buildings all have the effect of heating up the atmosphere, although probably not more than a very few degrees.

There is also some evidence of gradual warming of

the climate over larger areas. Since the Sierra Nevada glaciers were first measured by John Muir in the 1870's, they have receded a few feet, possibly indicating lighter snowfalls or warmer summers or both.

In many places on both the Pacific and Atlantic, sea-level measurements indicate a slow rise in the surface of the ocean at the rate of about a quarter of an inch a year. One possible cause is the melting of the polar ice, attributable to higher temperatures over a long period of time. But whether these phenomena are part of long-term climatic changes or simply periodic fluctuations no one can yet be sure; the statistics have not yet been accumulated over a sufficient period of time.

It is possible to "prove" almost anything by selecting evidence to fit a particular theory, but because of the relatively brief time during which reliable weather records have been collected, most meteorologists are skeptical of sweeping generalizations about long-term climatic changes. Studies of annular rings in California's redwoods indicate that in past centuries there have been both extended rainy periods and long droughts, each lasting as long as six years, but little evidence of regular cycles or long-term changes is discernible.

Fossil remains of plants and animals seem to indicate that in past epochs the weather of the portion of the earth now occupied by the San Francisco Bay Region has varied widely—from near-tropical to subarctic—but the cause and nature of these drastic changes are still unknown. In terms of any conceivable effect on present generations, these changes, consuming hundreds of thousands of years, have little significance.

Far more important are the changes that man himself may make. Cloud seeding is doubtless only the first of many steps to control or influence the weather. Changes in ocean currents by the construction of mammoth breakwaters offer other prospects for influencing the climate.

Probably the greatest man-made climatic effects in

the San Francisco Bay Region will be created inadvertently. Smog has already reduced drastically the purity and visibility of the atmosphere in the entire Bay Region, particularly within the mountain-surrounded basin of the bay itself. The gradual filling of the edges of the bay for building purposes is diminishing the bay's size and will similarly reduce the bay's effect on the climate of the shores.

Under natural conditions the bay acts as a kind of thermostat. On a far smaller scale than the ocean itself, but nevertheless to an important degree, the bay warms its shores in the winter and cools them in the summer. Refreshing breezes blow all summer from the cool bay to the warm land and lower the temperature of the shores very appreciably.

If the present filling process continues until all the shallows less than 18 feet deep are eliminated, the bay will shrink to about 25 per cent of its natural size, and the climate of the region around it will change proportionally. The bay is a vital element in the creation of the diverse microclimates that are an essential ingredient of the unique quality of life in the Bay Region. Like many other invaluable natural resources, San Francisco Bay will be largely destroyed by haphazard exploitation unless there are concerted regional efforts to plan its preservation and development on a rational basis.

SUGGESTED READING LIST

Blumenstock, David I. *The Ocean of Air*. New Brunswick: Rutgers University Press, 1959.

Brown, Slater. *World of the Wind*. Indianapolis: Bobbs-Merrill, 1961.

Cook, J. Gordon. *Our Astonishing Atmosphere*. New York: Dial, 1957.

De La Rue, E. Aubert. *Man and the Winds*. New York: Philosophical Library, 1955.

Fenton, Carroll Lane, and Mildred Adams Fenton. *Our Changing Weather*. New York: Doubleday, 1954.

Fisher, Robert Moore. *How about the Weather*. New York: Harper, 1958.

Flora, Snowden D. *Tornadoes of the United States*. Norman: University of Oklahoma Press, 1953.

Forrester, Frank. *1001 Questions Answered about the Weather*. New York: Dodd, Mead, 1957.

Hewitt, R. *From Earthquake, Fire and Flood*. New York: Scribner, 1958.

Kimble, George H. T. *Our American Weather*. Bloomington: Indiana University Press, 1955.

Kohn, Irving. *Meteorology for All*. New York: Barnes and Noble, 1946.

Krick, Irving P., and Roscoe Fleming. *Sun, Sea, and Sky*. New York: Lippincott, 1954.

Laird, Charles, and Ruth Laird. *Weathercasting*. Englewood Cliffs, N. J.: Prentice-Hall, 1955.

Longstreth, T. Morris. *Understanding the Weather*. New York: Macmillan, 1953.

Merit Badge Series. *Weather*. New York: Boy Scouts of America, 1943.

Patton, Clyde Perry. *Climatology of Summer Fogs in the San Francisco Bay Area*. Berkeley and Los Angeles: University of California Publications in Geography, Vol. 10 (3), 1956.

Scorer, R. S. *Weather*. New York: Roy, 1959.